What are the National Science Education Standards?

The National Research Council* released the *National Science Education Standards* in December of 1995. The *Standards* define the science content that all students should know and be able to do and provide guidelines for assessing the degree to which students have learned that content. The *Standards* detail the teaching strategies, professional development, and support necessary to deliver high quality science education to all students. The *Standards* also describe policies needed to bring coordination, consistency, and coherence to science education programs.

The National Science Education Standards *include standards for*

- Content
- Teaching
- Assessment
- Professional Development
- Program
- System

*The National Research Council is the principal operating agency of the National Academy of Sciences, the National Academy of Engineering, and the Institute of Medicine.

Why do we need the Standards?

- Understanding science offers personal fulfillment and excitement.
- Citizens need scientific information and scientific ways of thinking in order to make informed decisions.
- Business and industry need entry-level workers with the ability to learn, reason, think creatively, make decisions, and solve problems.
- Strong science and mathematics education can help our nation and individual citizens improve and maintain their economic productivity.

Who developed the Standards?

Committees and working groups of scientists, teachers, and other educators appointed by the National Research Council developed the *Standards*. They engaged in a four-year process that involved review and critique by 22 science education and scientific organizations and broad state and local participation of over 18,000 individuals, including scientists, science educators, teachers, school administrators, and parents. The national consensus that resulted from this process gives the *Standards* a special credibility. Educators throughout the country who use them to inform changes in science education programs can be assured that the *Standards* represent the highest quality thinking this country can provide its citizens.

The vision of the Standards:

All students, regardless of age, gender, cultural or ethnic background, disabilities, aspirations, or interest and motivation in science, should have the opportunity to attain high levels of scientific literacy.

Guiding Principles behind the Standards

- **Science is for all students.**
- **Learning science is an active process.**
- **School science reflects traditions of contemporary science.**
- **Improving science is part of systemwide educational reform.**

How do students learn science?

The *Standards* are based on the premise that learning science is something that students do, not something that is done to them. The *Standards* envision an active learning process in which students describe objects and events, ask questions, formulate explanations, test those explanations, and communicate their ideas to others. In this way, students build strong knowledge of science content, apply that knowledge to new problems, learn how to communicate clearly, and build critical and logical thinking skills.

Through their study of science, students

- **Experience the richness and excitement of the natural world**
- **Apply scientific principles and processes to make personal decisions**
- **Discuss matters of scientific and technological concern**
- **Increase their potential contribution to society and to the economy**

What should students know and be able to do?

The **Content Standards** describe the knowledge and abilities students need to develop, from kindergarten through high school, in order to become scientifically literate.

What is scientific literacy?

Scientific literacy is the knowledge and understanding of scientific concepts and processes required for personal decision making, participation in civic and cultural affairs, and economic productivity. People who are scientifically literate can ask, find, or determine answers to questions about everyday experiences. They are able to describe, explain, and predict natural phenomena.

Scientific literacy has different degrees and forms; it expands and deepens over a lifetime, not just during the years in school. The *Standards* outline a broad base of knowledge and skills for a lifetime of continued development in scientific literacy for every citizen, as well as provide a foundation for those aspiring to scientific careers.

How are the National Science Education Standards different from the American Association for the Advancement of Science's Benchmarks for Science Literacy?

The documents differ in three ways. First, they divide content by different grade levels. The *Benchmarks* are statements of what all students should know and be able to do in science, mathematics, and technology by the end of grades 2, 5, 8, and 12; the *Standards* use grades 4, 8, and 12 as end points. Second, the *Standards* place greater emphasis on inquiry, including it as important science content as well as a means of teaching and learning. Third, the *Standards* offer a broader set of standards for improving science education. They address all components of education, including teaching, assessment, professional development, program, and system, recognizing that improvement cannot occur or be sustained in one segment of the system alone. There is, however, a high level of consistency between the two documents in describing the content to be learned. The National Research Council believes that the use of the *Benchmarks* complies fully with the spirit of the content standards.

What is included in content standards?

Content standards are divided into eight categories:

- **Unifying concepts and processes**
- **Science as inquiry**
- **Physical science**
- **Life science**
- **Earth and space science**
- **Science and technology**
- **Science in personal and social perspectives**
- **History and nature of science**

The content standards include traditional school science content but, in addition, encompass other knowledge and abilities of scientists. The first category of the content standards, unifying concepts and processes, identifies powerful ideas that are basic to the science disciplines and help students of all ages understand the natural world. This category is presented for all grade levels because the concepts are developed throughout a student's education. The other content categories are clustered for grades K-4, 5-8, and 9-12. Students develop knowledge and abilities in inquiry, which ground their learning of subject matter in physical, life, and earth and space sciences. Science and technology standards link the natural and designed worlds. The personal and social perspectives standards help students see the personal and social impacts of science and help them develop decision-making skills. The history and nature of science standards help students see science as a human experience that is on-going and ever-changing.

CONTENT STANDARDS

	Grades K-4	Grades 5-8	Grades 9-12
Unifying Concepts and Processes	Systems, order, and organization Evidence, models, and explanation Change, constancy, and measurement Evolution and equilibrium Form and function	Systems, order, and organization Evidence, models, and explanation Change, constancy, and measurement Evolution and equilibrium Form and function	Systems, order, and organization Evidence, models, and explanation Change, constancy, and measurement Evolution and equilibrium Form and function
Science as Inquiry	Abilities necessary to do scientific inquiry Understandings about scientific inquiry	Abilities necessary to do scientific inquiry Understandings about scientific inquiry	Abilities necessary to do scientific inquiry Understandings about scientific inquiry
Physical Science	Properties of objects and materials Position and motion of objects Light, heat, electricity, and magnetism	Properties and changes of properties in matter Motions and forces Transfer of energy	Structure of atoms Structure and properties of matter Chemical reactions Motions and forces Conservation of energy and increase in disorder Interactions of energy and matter
Life Science	Characteristics of organisms Life cycles of organisms Organisms and environments	Structure and function in living systems Reproduction and heredity Regulation and behavior Populations and ecosystems Diversity and adaptations of organisms	The cell Molecular basis of heredity Biological evolution Interdependence of organisms Matter, energy, and organization in living systems

Earth and Space Science	Properties of earth materials Objects in the sky Changes in earth and sky	Structure of the earth system Earth's history Earth in the solar system	Energy in the earth system Geochemical cycles Origin and evolution of the earth system Origin and evolution of the universe
Science and Technology	Abilities of technological design Understandings about science and technology Abilities to distinguish between natural objects and objects made by humans	Abilities of technological design Understandings about science and technology	Abilities of technological design Understandings about science and technology
Science in Personal and Social Perspectives	Personal health Characteristics and changes in populations Types of resources Changes in environments Science and technology in local challenges	Personal health Populations, resources, and environments Natural hazards Risks and benefits Science and technology in society	Personal and community health Population growth Natural resources Environmental quality Natural and human-induced hazards Science and technology in local, national, and global challenges
History and Nature of Science	Science as a human endeavor	Science as a human endeavor Nature of science History of science	Science as a human endeavor Nature of scientific knowledge Historical perspectives

What do teachers of science do?

The **Teaching Standards** provide an answer to this question. Science teaching lies at the heart of the vision of science education presented in the *Standards*. Effective teachers of science have theoretical and practical knowledge about student learning, science, and science teaching. The teaching standards describe actions these teachers take and skills and knowledge they have to teach science well.

Teachers of science

- **Plan an inquiry-based science program**
- **Guide and facilitate learning**
- **Assess student learning and their own teaching**
- **Design and manage learning environments**
- **Develop communities of science learners**
- **Participate in on-going development of the school science program**

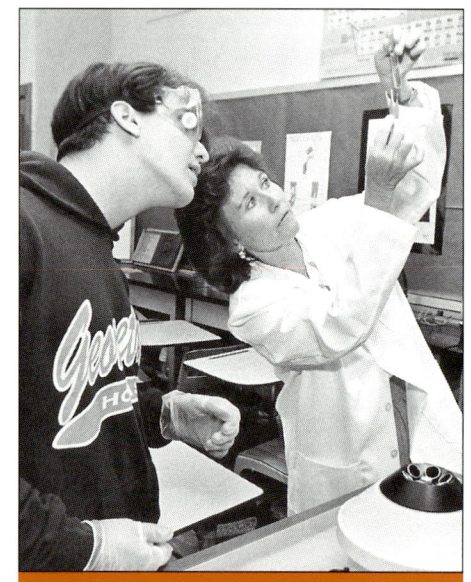

How can teachers apply the Standards in their classrooms?

Individual teachers are encouraged by the *Standards* to give less emphasis to fact-based programs and greater emphasis to inquiry-based programs that engage students in an in-depth study of fewer topics. However, to attain the vision of science education described in the *Standards*, more than teaching practices and materials must change. The routines, rewards, structures, and expectations of districts, schools, and other parts of the system must endorse the vision, and provide teachers with resources, time, and opportunities to change their practice. Teachers can use the program and system standards to communicate this need to administrators and parents.

How is science learning assessed?

The **Assessment Standards** provide criteria to judge progress across the system toward the science education vision of scientific literacy for all. They can be used in preparing evaluations of students, teachers, programs, and policies.

Assessments should

- **Be deliberately designed for the decisions they are intended to inform**
- **Measure both achievement and opportunity to learn**
- **Clearly relate decisions to data**
- **Demonstrate fairness in design and use**
- **Support their inferences with data**

Will the Standards *help teachers test their students more effectively?*

Teaching and testing are integral components of instruction, and cannot be separated. As content and teaching strategies become aligned with the *Standards*, so must classroom assessments. The assessment standards identify essential characteristics of effective assessment policies, practices, and tasks at all levels. Teachers who use the standards will think differently about what to assess, when to do so, and the best ways to determine what their students are learning. They will consider carefully the fundamental understandings their students are working to learn, the place their students are in developing understanding, and a variety of alternatives to help their students demonstrate what they know.

Will standardized tests change?
The *Standards* address the need for systems to reconsider the purpose, data analysis, and sample size in all large-scale assessments. There are already indications that changes in items on common standardized tests are being considered, as are the designs used by states, districts, and others who conduct large-scale science assessments.

What do teachers need to know and how will they learn it?

The **Professional Development Standards** make the case that becoming an effective teacher of science is a continuous process, stretching from preservice throughout one's professional career. The professional development standards can be used to help teachers of K-12 science have the on-going, in-depth kinds of learning opportunities that are required by and available to all professionals.

Professional Development Standards call for teachers to have opportunities to

- Learn science through inquiry
- Integrate knowledge of science, learning, and teaching
- Engage in continuous reflection and improvement
- Build coherent, coordinated programs for professional learning

How will teachers gain the science content knowledge they need? To help their students achieve high levels of science literacy, teachers need to understand deeply the content they teach. Building science knowledge

- **Involves active investigation**
- **Focuses on significant science**
- **Uses scientific literature and technology**
- **Builds on teachers' current knowledge**
- **Encourages on-going reflection**
- **Supports collaboration among teachers**

How will teachers improve their science teaching? Effective teachers of science have specialized knowledge that combines their understanding of science with what they know about learning, teaching, curriculum, and students. They develop this unique type of knowledge through both preservice and inservice learning experiences that

- **Deliberately connect science and pedagogy**
- **Model effective teaching practices**
- **Address the needs of teachers as adult learners**
- **Take place in classrooms and other learning situations**
- **Use inquiry, reflection, research, modeling, and guided practice**

What is an effective school science program?

The **Program Standards** address the need for comprehensive and coordinated science experiences across grade levels and support needed by teachers in order for all students to have opportunities to learn. The program standards will help schools and districts translate the *Standards* into effective programs that reflect local contexts and policies.

Program Standards call for

- **Consistency across all elements of the science program and across K-12**
- **Quality in the program of studies**
- **Coordination with mathematics**
- **Quality resources–teachers, time, materials**
- **Equitable opportunities for achievement**
- **Collaboration within the school community to support a quality program**

Quality Programs of Study

- **Include all content standards**
- **Select developmentally appropriate content**
- **Emphasize student understanding through inquiry**
- **Connect science to other subjects**

Are the **Standards** *a science curriculum?*

Curriculum is the way content is designed and delivered. It includes the structure, organization, balance, and presentation of the content in the classroom. The *Standards* do not prescribe a specific curriculum but, rather, provide criteria that can be used at the local, state, and national levels to design a curriculum framework, a key element in a school or district's science program, or to evaluate and select curriculum materials. Effective science programs are designed to consider and draw consistency from the content, teaching, and assessment standards, as well as professional development, program, and system standards.

How does the system support science learning?

The **System Standards** call on all parts of the educational system—including local districts, state departments of education, and the federal education system—to coordinate their efforts and build on one another's strengths. The standards can serve as criteria for judging the performance of components of the system responsible for providing schools with necessary financial and intellectual resources.

System Standards require

- **Policies consistent with vision of the Standards**
- **Coordination of policies within and across system**
- **Continuity of support over time**
- **Sufficient resources to support program**
- **Equitable policies**
- **Attention to anticipated effects**
- **Individual responsibility for achieving the vision**